THE LARGEST INDOOR AND OUTDOOR MARIJUANA FARMS IN THE WORLD

MICKEY DEE

Frazier Publishing & Services

P. O. Box 363835

North Las Vegas, NV 89036

Table of Contents

INTRODUCTION

Starting with the soil!

Soil is required, except for cannabis grown with hydroponics or aeroponics.

Sufficient nutrients—commercial potting soils usually indicate this as "N-P-K = x%-y%-z%" the percentages of the fundamental nutritional elements, i.e., nitrogen, phosphorus and potassium. Nutrients are often provided to the soil via fertilizers but such practice requires caution. pH between 5.8 and 6.5. This value can be adjusted

– see soil pH. Commercial fertilizers (even organic) tend to make the soil more acidic.

Warmth

The optimal day temperature range for cannabis is 24 to 30 °C (75 to 86 °F). Temperatures above 31 °C (88F) and below 15.5 °C (60F) seem to decrease THC potency and slow growth. At 13 °C (55F) the plant undergoes a mild shock, though some strains withstand frost temporarily.

Light

Light can be natural (outdoor growing) or artificial (indoor growing).

Under artificial light, the plant typically remains under a regime of 16–24 hours of light and 0–8 hours of darkness from the germination until flowering, with longer light periods being conducive to vegetative growth, and longer dark periods being conducive to flowering. However, generally Cannabis only requires thirteen hours of continuous light to remain in the vegetative stage. The 'Gas Lantern Routine' is an alternate lighting schedule that has proven to be successful for growing Cannabis, while saving a significant amount of energy. For optimal health, Cannabis plants require a period of light and a period of dark. It has been suggested that, when subjected to a regimen of constant light without a dark period, cannabis begins to show signs of decreased photosynthetic response, lack of vigor, and an overall decrease in vascular development. Typically, flowering is induced by providing at least 12 hours per day of complete darkness. Flowering in cannabis is triggered by a hormonal reaction within the plant that is initiated by an increase in length of its dark cycle, i.e. the plant needs sufficient prolonged darkness for bract/bracteole (flowering) to begin. Some Indica varieties require as little as 8 hours of dark to begin flowering, whereas some Sativa varieties require up to 13 hours.

Water

Watering frequency and amount is determined by many factors, including temperature and light, the age, size and stage of growth of the plant and the medium's ability to retain water. A conspicuous sign of water problems is the wilting of leaves. Giving too much water can kill cannabis plants if the growing medium gets over-saturated. This is mainly due to oxygen not being able to enter the root system.Anaerobic bacteria start to accumulate due to waterlogged, stale conditions. They begin to consume plant roots, beneficial(aerobic) bacteria, as well as nutrients and fertilizer. When using soil as a growth medium, the soil should be allowed to dry adequately before re-watering.

Humidity

Humidity is an important part of plant growth. Dry conditions slow the rate of photosynthesis. Ideal levels of humidity for optimal growth are forty to sixty percent.

Nutrients

Nutrients are taken up from the soil by roots. Nutrient soil amendments (fertilizers) are added when the soil nutrients are depleted. Fertilizers can be chemical or organic, liquid or powder, and usually contain a mixture of ingredients. Commercial fertilizers indicate the levels of NPK (nitrogen, phosphorus, and potassium). In general, cannabis needs more N than P and K during all life phases. The presence of secondary nutrients (calcium, magnesium, sulfur) is recommended. Micro nutrients (e.g. iron, boron, chlorine, manganese, copper, zinc, molybdenum) rarely manifest as deficiencies.

Because Cannabis' nutrient needs vary widely depending on the variety, they are usually determined by trial and error and fertilizers are applied sparingly to avoid burning the plant.

WORLD'S LARGEST INDOOR CANNABIS FACILITY – FULLY BACKED & GROWING

The legalization of recreational cannabis is getting closer every day in Canada and producers are busier than Santa's elves on Christmas Eve making sure their operations are ready to provide the provinces and territories with enough supply.

Will it be enough? Some journalists don't think so. However, one Company in the business is planning to produce upwards of 400,000 kg of pot a year from the biggest production facility on earth.

FSD Pharma Inc. (CSE:HUGE, Forum) is working to target all of the legal aspects of the medicinal cannabis industry under one roof, from cultivation, to manufacturing, extracts and research and development.

FSD is building out the world's largest indoor hydroponic cultivation and processing facility. The Company purchased the 630,000 sq. ft. facility in Cobourg, Ontario from KRAFT Foods (NYSE:KFT), who previously used the building as a food manufacturing facility. Sitting on 70 acres of land, it has 40 acres primed for development. It has the potential to host more than 3.8 million sq. ft. of cultivation and processing area. FSD purchased this facility through its wholly- owned subsidiary FV Pharma Inc., a Licensed Producer under the Access to Cannabis for Medical Purposes Regulations (ACMPR).

What makes this facility unique is not only the size, but its focus on indoor hydroponic cannabis only, no greenhouses or outdoor operations. Cannabis growing is ultimately an exercise in agriculture, as well as an art.

Expanding upon FSD's output power through FV Pharma, the Company recently signed a deal with Cannara Biotech Inc. where it will occupy more than 105,000 square feet of Cannara's 625,000 square foot facility, located 45 minutes fromdowntown Montreal.

FSD owns a 25% share in Cannara.

The challenge with growing is that varied conditions result in varied products. This means that one producer with different facilities wouldn't necessarily produce the same product from one crop to another in a different location. Aurora Cannabis Inc.

(TSX: ACB, OTCQB: ACBFF) for example, has several facilities across several provinces, which means different grow teams, different standard operating procedures and nuances between crops. This results in an end product that is different depending on where it comes from. For economy of scale and cost efficiency, having everything under one roof controls the product and controls costs.

Another key factor that makes indoor growing important is that most producers are looking to grow cannabis flowers in greenhouses. Growing indoor not only avoids that competition, but targets consumer demand for a better product. In both the U.S. and Canada, the highest prices per-gram on average is paid for cannabis grown indoor, not in greenhouses. This is due in-part because growing indoors is more expensive, but slashing expenses to do so provides more consistency. A greenhouse still relies on the sun for growing, which gives different environments throughout the year. In turn, this can affect the soil and terpene profiles. Consumers also find that their tolerances can often grow more quickly with greenhouse products than those grown indoors. Operating such a large facility also means that there is a lot of flexibility where the sheer size means FV Pharma can test different plants across multiple rooms.

FV already has access to premium cannabinoid-based products and scientific research developed by Israel-based firm SciCann Therapeutics Inc. through a $3 million investment for a 15% equity stake. SciCann's platform offers a series of rigorous clinical studies through a network of leading researchers, academic institutions and medical centres. FV has received an exclusive license in Canada to produce and distribute SciCann's line of products. In focusing on the pharmaceutical medicinal side of the business, FV Pharma's strain variations will focus on entourage profiles, such as CBD's, CBN's, THC's and CBG's, to determine exactly which conditions yield the best results to what the Company is looking for. The environment of growing conditions can often dictate how growing will react. This is why HUGE is targeting the medicinal market first, because of its high standards. The recreational market will always fall in line to those high standards. The Company is striving to be at the forefront of testing and using different types of technologies for growing and with such a large facility they can determine which systems offer the more effective advantages.

Anthony Durkacz, the Executive Vice President of First Republic Capital and Director for HUGE spoke to Stockhouse Editorial about the Company's vision for this facility becoming a "Cannabis Wonderland", where it could manufacture everything from

flowers, oils, pills, vaporizers, beverages, edibles, but will always keep its focus on the pharmaceutical side.

Location,

Along with its massive square footage, the facility's location is another great asset. The site is optimally situated just off Highway 401, about an hour east of Toronto, Canada's largest investing community. The next nearest cannabis producing facilities in the area are more than three hours away. Investors can visit this site with ease and Toronto consumers can receive medical marijuana products within the same day. FSD is focusing on rolling out its products in Ontario but is looking to expand nationally and internationally. Any legal market in this space the Company can sell to for the right price, it will. In addition to its cultivation license, FSD also has import / export licenses.

On-site, it has an electrical sub-station, natural gas lines, multiple water intakes, as well as rail lines that feed directly into the facility with 26 loading docks. Once this facility is operating at full capacity, it is expected that FV Pharma will receive around 400 million grams of dried cannabis flower a year, of which it will take 50.1%.

HUGE signed a strategic partnership deal with Auxly Cannabis Group Inc. (TSX: V.XLY, OTCQB: CBWTF, Forum) (Formerly Cannabis Wheaton Income Corp.) back in December 2018.

Auxly Cannabis will finance, build and develop all aspects of the cultivation facility in stages (except the area already built and growing by HUGE) and

Auxly will endure all of the costs, while HUGE operates the grow operations through it's wholly owned subsidiary FV Pharma. Specific to areas that HUGE's partner Auxly

Cannabis Group Inc. finances and builds for all cannabis and cannabis-related production, HUGE will retain all operating costs plus ten percent profit and then anything left over will be split 50.1% to HUGE and 49.9% to Auxly. Not only does HUGE own the entire former Kraft property with no debt, the Auxly deal is a significant source of financing for the Company's operations and is a major source of stability, currently building out the facility's grow rooms to the point of operation, meaning the Company won't have to bear those costs. HUGE also raised approx. $53 million through private placements by selling shares without warrants, which also means less dilution as the share price climbs. Capital costs for such ambitions are high, this relationship means it won't have to continuously go to market for capital.

A major reason for the XLY deal revolves around its chairman and CEO, Chuck Rifici.

The marijuana industry pioneer was also the cofounder of major producer Canopy Growth Corp. (T.WEED) when that company bought the former Hershey chocolate plant in Smiths Falls, Ontario. He brings more experience than nearly anyone at converting a large food-grade facility into a successful indoor grow operation.

The facility is currently growing in four rooms. HUGE, along with Auxly Cannabis have applied to Health Canada to break ground on a 200,000 sq. ft. expansion. XLY also recently completed a bought deal for $100 million in financing, which will give a helping hand toward further expansion.

Generating Unprecedented Penny Stock Volumes

Click to enlargeThis helped the Company pave the way to trade on the Canadian Securities Exchange, but it didn't stop there. On its first day of trading, FSD broke the all-time daily volume record with more than 78 million Class B subordinate voting shares. It followed this success up by breaking the all-time weekly volume record a few days later when it recorded over 259 million Class B subordinate voting shares. FSD's trading volume was nearly 50% bigger than the previous record holder.

The company has since broken its own all-time weekly volume record and recorded over 269 million Class B subordinate voting shares.

Company President and CEO Thomas Fairfull laid out the benefit of publicly trading on the CSE for investors:

"From inception to where the Company has grown today is tremendously exciting and I wish to thank the world-class team we have assembled and the investors that have supported our vision. The Canadian cannabis industry is developing with pace and we believe quality cannabis will be the backbone of all our future cannabis, cannabis products, ancillary business and pharmaceutical development initiatives. As such, we are committed to cultivating the highest quality cannabis at scale by building out the world's largest, state of the art, hydroponic cultivation and processing facility, which will support all our planned business units under one massive roof.

Going forward, we will continue to aggressively pursue our business plan and update our shareholder community in an ultimate effort to deliver significant shareholder value."

Investors looking to diversify their portfolio with a marijuana producer should do their due diligence and give FSD Pharma Inc. a closer look. The Company is in business already, with an ambitious plan to be the largest producer, already breaking trading records with secure funding in place to fully take advantage of all aspects of the cannabis industry from seed to sale.

WORLD'S TOP CANNABIS PRODUCING COUNTRIES

Many Americans purchase their weed directly (and sometimes covertly) from a single, intermediary supplier. Although this is an efficient system, it leaves the consumer clueless about the product's supply chain. The artisan movement in food has provoked a lot of interest about the origins of what we eat — we can only hope that the gradual legalization of cannabis will lead more people to be informed about the provenance of their cannabis as well.

Whether native or introduced over centuries of human communication, cannabis has thrived in a variety of areas and climates around the world. Here's a roundup of some of the top cannabis-producing countries in the world, and a look at the ways their output has been affected by trade and policy.

Colombia

Although Colombia is often caricatured as a center of cocaine production, it also plays a large role in the growth of high-grade cannabis. The plant was introduced to Colombia from Panama in the 1920s and took well to the temperate climate of its mountain regions — so well, in fact, that by the 1970s and 80s, Colombian cannabis accounted for 70% of all cannabis smuggled into the U.S.

Colombia legalized medical cannabis in December 2015 and showed some early indication of encouraging its cultivation among local farmers; however, a popular referendum to reject a government peace deal with the terrorist militia group FARC may mean that FARC will remain in illicit control of rural cannabis in Colombia for some time to come.

Mexico

Mexico gave us the word "marijuana," so it shouldn't be a surprise that the country has historically been one of the largest cannabis producers in the world. Indoor mexican grow op-Potent sativa varieties like Acapulco Gold grown in Mexico's humid coastal regions became especially popular with American smokers in the 1960s and 70s and, as recently as 2008, Mexican cannabis accounted for two-thirds of all cannabis consumed in the United States. Recent legalization in the U.S. has put a crimp in the Mexican market, though — homegrown American weed may have decreased production in Mexico by up to 30%, leading the country's cartels to begin pushing harder drugs across the border.

India

India is a point of origin for the variety of cannabis we know today as indica (indica is an outdated latinized term referring to the Indian subcontinent).

Indica's short, bushy plants may have been cultivated by early Indian cannabis farmers for easy manual access to resin which they used to make an ancient hash-like concentrate called charas. Although not widely exported, Indian cannabis is still grown in large quantities for local use in the production of state-sanctioned charas and a THC milkshake-like preparation called bhang.

Afghanistan

Along with India, Afghanistan is considered a birthplace of modern indica varieties.

The country's cannabis legacy made it a popular hippie destination in the 1960s and 70s, and resulted in the importation of Afghani landrace strains into the Outdoor grow opU.S. for crossbreeding. Although the Nixon administration pressured the Afghan government to ban all cultivation of the plant, cannabis has since made a resurgence; in 2010, the country was the world's largest cannabis supplier, growing between 25,000 and 59,000 acres every year. As in Colombia, cannabis cultivation in Afghanistan has unfortunately also became a source of revenue for a terrorist group: the Taliban has derived significant income by taxing the growth and trafficking of illicit cannabis.

South Africa

Just as its place at the tip of the African continent has made South Africa a center of biodiversity, its unique climate has also made the country particularly conducive to growing cannabis. Possibly introduced to the continent by Arab traders in the Middle Ages, sativa varieties thrived in the country's mountainous coastal regions and were used to crossbreed popular strains like Durban Poison. Cannabis has become an extremely successful export crop for the country — a 2004 report estimated that 99% of all cannabis consumed in Ireland was grown in South Africa.

The relative ease of trafficking cannabis from South Africa's extensive coastline has even let its bordering countries to focus their agricultural efforts on cannabis production for export to South Africa.

Jamaica

Cannabis was brought to Jamaica in the 19th century by Indian indentured servants traveling in tow with English colonists. Yet despite its long association with the island, cannabis has been illegal in Jamaica for the last several decades.Jamaican street market Jamaica is the largest Caribbean supplier of illicit cannabis to the United States and to other Caribbean islands. However, following the success of gradual legalization in the U.S., the Jamaican government decriminalized cannabis use and cultivation in 2015 and made the plant legal for adherents of the Rastafari religion. This relaxed attitude has allowed for some renewed focus on the country's high-quality sativas: in 2015, High Times magazine hosted an offshoot of its annual Cannabis Cup in Negril, with a second event planned for 2016.

Netherlands

During the 1980s and 90s, The Netherlands were a haven for not only cannabis consumption, but also advanced cultivation. The country's lax attitude towards cannabis regulation during this period attracted advanced botanists who preserved many heritage strains and founded seed companies like Sensi and Dutch Passion.

Although the government still tolerates consumption in designated Amsterdam coffeehouses, it has since cracked down on personal and commercial cultivation of cannabis. Nevertheless, The Netherlands has been a cornerstone of refined crossbreeding.

Kazakhstan

Nestled in Central Asia, Kazakstan has a long history as a major grain supplier for much of the surrounding region. It's also a hotbed of cannabis cultivation.

The plant grows wild across the country, but particularly in its fertile Chuy Valley where low-THC but high quality cannabis is valued by both locals and consumers in neighboring Russia. Although Soviet forces attempted to eradicate the plants in the 1980s, their efforts were fortunately unsuccessful. Kazakhstan is also a central location for processing of hashish, which is cultivated by riding horses through cannabis fields and then scraping the resin off of their bodies.

THE 6 MOST ADVANCED COUNTRIES IN TERMS OF MEDICAL CANNABIS RESEARCH

Israel is the country that pioneered cannabis research, and it still remains a major hub for research in the field.

Canada, the Czech Republic and Spain are also among the countries engaged in researching cannabis and its effects.

Keep scrolling to find out more about the countries that are making efforts to bring cannabis closer to patients.

Multiple sclerosis, epilepsy, schizophrenia, Alzheimer's, fibromyalgia, chronic pain... The therapeutic effects of cannabis make the plant a great treatment option for a number of conditions. This is probably the reason why an ever-increasing number of international researchers are exploring its medical benefits.

Here are the countries that are contributing the most to research on therapeutic cannabis.

Israel, the initiator

Besides being a world-leader in medical cannabis research, Israel is the country that pioneered the study of the chemical properties of the substance. In fact, it was Raphael Mechoulam, professor and researcher at the Hebrew University of Jerusalem, that first isolated THC more than 50 years ago. And not just that, he also synthesized cannabidiol (CBD), carrying out the first study into its effects on epilepsy patients over three decades ago.

Mechoulam has been devoted to research ever since, and only a few months ago embarked on yet another project, this time as the leader of a team responsible for researching theeffects of CBD on asthma patients.

Israel started promoting research in this field over a decade ago, when the Government launched an ambitious programme aimed at boosting medical cannabis at national level, providing access to prescription cannabis to over 25,000 Israelis and encouraging scientists, producers and institutions to engage more actively in the research and development of the plant.

Canada, the country of big companies

Canada's Prime Minister Justin Trudeau is set to legalise recreational marijuana within the summer. Plus, his Government has committed to support 14 research projects focusing on cannabis through funding from the Canadian Institutes of Health Research.

This, however, is hardly surprising, as Canada is among the countries that pioneered a more liberal approach to cannabis: in the late '90s the Government implemented a programmed aimed at granting access to medical cannabis and, in the last few years, the regulation has been modified to cover the cultivation of cannabis for medical purposes in an effort to make it easier for companies to get in the market. As a result, a number of Canadian companies have become industry giants in terms of research, production and exports.

A number of Canadian companies have become industry giants in terms of research.

The Czech Republic, at the cutting edge of research

The central European country regulated medical cannabis in 2013 so that patients including cancer and chronic pain sufferers could have prescription access to the substance. And while high prices are not helping consumption, it is also true that the country is home to one of the most cutting-edge research centres in the field of medical cannabis: the International Cannabis and Cannabinoids Institute (ICCI).

Located in Prague and opened at the end of 2015 with backing from American organisations and from the Czech Ministry of Health, amongst others, this multidisciplinary centre works with universities, tech companies and organisations from around the world which have an interest in developing therapeutic cannabis. In fact, the purpose of the ICCI is "to enable scientific examination of the relation between bioactive cannabis compounds and the effect on the human organism in the treatment of specific syndromes."

Spain, land of senior researchers

Home cultivation is permitted in Spain as long as it is not for selling purposes.

Accordingly, the Spanish Agency of Medicines and Medical Devices (AEMPS) may issue cultivation licenses for therapeutic and research purposes, with five companies having already obtained a permit and benefiting from 20,000 hectares where they can grow legally.

Meanwhile, various Spanish research groups have made significant progress in the last few years. At the Complutense University of Madrid, for instance, a team of scientists led by Dr Guillermo Velasco has been researching the applications of cannabinoids for the treatment of various diseases for over a decade, and as early as 1998 researchers from the same university discovered that THC may induce programmed cell death of tumour cells. Moreover, a research group led by Dr Manuel Guzmán announced in 2002 that they had used THC to destroy incurable brain tumours in rats.

Various Spanish research groups have made significant progress in the last few years.

A few years later, in 2015, top experts from the field of cannabis research, monitoring and outreach founded the Spanish Observatory on Medical Cannabis with a view to promoting, coordinating and organizing activities aimed at increasing awareness of the therapeutic proper-ties of cannabis and its derivatives.

5- The Netherlands, slow but steady

In 2003, The Netherlands allowed the sale of medical cannabis in pharmacies to qualifying patients. And while research is also permitted in the country, it is subject to strict regulation, with the Office of Medical Cannabis – run by the Ministry of Health – being fully responsible for the production and distribution of the cannabis to universities, pharmacies and research centres.

In the same vein, Bedrocan is the sole producer supplying medical cannabis in the country. Besides conducting its own studies, the company collaborates with other research centres, including the Leiden University Medical Center, which is currently researching the effects of inhaled cannabis on the symptoms of fibromyalgia.

6– Uruguay, the next research hub?

Over and above these countries, there are other nations that seem set to become major hubs of research in the years to come. One of them is definitely Uruguay, the first country in the world to have fully legalised the production and selling of cannabis nationwide, along with the export of medical cannabis.

Earlier this year, the country's president, Tabaré Vázquez, announced the opening of a large, privately-owned cannabis research and production plant, but despite the favourable regulatory environment, researchers across the country complain that the lack of funding is affecting their research projects.

Another South American country that could become research friendly in the near future is Colombia. The cultivation of cannabis for medical and scientific purposes is already legal, and researchers across the world regard the country as an ideal location because of its favourable weather and its low production costs.

An interesting case is that of the United States. While many of the studies on therapeutic cannabis are conducted there - 29 states have already legalised the medical use of cannabis - the substance is not legal federally, meaning that researchers have to overcome a myriad of obstacles in what is an administrative battle that may last for years before they can conduct one clinical study.

Accordingly, the United States is not exactly what one would describe as a friendly environment for cannabis research. Presumably, though, the list of countries interested in making the benefits of cannabis available to its citizens will just keep growing in coming years.

Differences between cannabis and hemp

Cannabis and hemp are two plants that often lead to confusion. Despite belonging to the same genus, their morphological differences and the...

Cannabis: Differences between Feminised and Autoflowering Seeds If you're thinking about starting to grow cannabis, you may find yourself immersed in a sea of doubt. What culture system shall I use? What.

A CANNABIS-BUSINESS PARK COVERING 1 MILLION SQUARE FEET

California is at the forefront of the US medical marijuana industry, and weed's positive impact on the state's economy has been huge, generating $2.8 billion in 2015, with $6.5 billion annually expected by 2020.

The state's largest grow facility was announced in June 2016 by GFarms, and is slated to be built in the town of Desert Hot Springs, which declared itself insolvent in 2014 and is now experiencing a real estate boom thanks to the marijuana industry. The GFarms facility will be 100,000 square feet and consist of three greenhouses on seven acres.

But it's going to be eclipsed in size before long.

AmeriCann, a Colorado company, has announced much bigger plans to build the nation's largest marijuana grow facility—in Massachusetts—in 2017.

Obviously, the prospects of enjoying an influx of cannabis cash similar to California's is appealing to other states and legalization proved popular in the November elections. The national marijuana market is projected to generate $50 billion a year by 2026. But the transition from underground illegal drug trade to legitimate business isn't fast or easy.

Large-scale projects are few and far between. It's difficult to get financing to go into a business that is still illegal federally, so big marijuana projects—while potentially profitable—are shirked by the corporations most likely to be interested in this new industry; Big Tobacco and Big Pharma aren't transforming into Big Pot yet.

There isn't some megalithic industry that exists today. There's no Philip Morris, no Anheuser-Busch, no cannabis division at Bank of America. Even the most successful company is still barely in the growth stage," Kris Krane, president of Massachusetts marijuana investment and consulting firm 4Front Ventures, told The Boston Globe (paywall).

The weed business is still very iffy—apart from financing problems, you can't transport cannabis across states because of its status as a Schedule 1 drug, and there are numerous obstacles to entry, such as obtaining one of the limited number of licenses available in any given state. "It's not for the faint of heart," says Tim Keogh, president and CEO of AmeriCann.

Keogh became interested in cannabis in 2010 when a friend dying of stomach cancer in Florida, where medical marijuana was unavailable, mentioned its therapeutic effects.

They joked about Keogh going out on the street to buy some but, in all seriousness, he was concerned about molds, potency, and other potential dangers associated with illegal drugs (like arrest, presumably). After his friend passed away, Keogh moved to Rhode Island, where medical marijuana was (and still is) legal, and started volunteering at another friend's dispensary.

He became a marijuana legalization advocate in 2011, and in 2012 joined the Board of Coastal Compassion, a Massachusetts dispensary he eventually ran. That same year, Keogh founded the Cannabis Producers Association of New England, an advisory board, and in 2014 joined AmeriCann, an early entrant in the medical marijuana market in Colorado, as president.

This year, his experience in New England helped him facilitate the purchase of a Massachusetts property expected to be the largest and most technologically advanced grow facility in the country. In March of 2017, construction is slated to begin on a 53-acre tract in Freetown, originally intended to be a Boston Beer Company brewery and acquired by AmeriCann this fall for $4.475 million.

The canna-business park will be 1 million sq ft and include energy-efficient greenhouses for cultivation, plant processing spaces, facilities for creating infused products, a testing laboratory, research and training centers, and corporate offices. Space will be sold or leased to businesses registered under the Massachusetts Medical Marijuana Program, and AmeriCann has a Host Community Agreement from Freetown that will enable businesses in the park to get streamlined preferential licensing. AmeriCann said in a statement that it "will set a new cannabis industry standard for energy efficiency, cost control, clean cultivation practices, and the production of Nutraceutical-grade infused products for the patients of Massachusetts."

Medical marijuana has been legal in Massachusetts since 2012, but recreational weed just became legal in November. Keogh told the Boston Business Journal the November vote supercharged the plans for Freetown; development of the facility is expected to be completed faster to accommodate the growing business opportunities and AmeriCann is eying other locations.

The marijuana investment firm Arcview Group projects that cannabis will bring $1 billion annually to Massachusetts by 2020. Douglas Leighton, co-founder of Boston- based hedge fund Dutchess Capital, is betting big on canna-biz startups—investing in 20 such companies between 2103 and 2016—before federal laws change and powerful corporate interests get in and start profiting from weed at the expense of smaller companies. He told Boston magazine,

"Eventually that will happen. Because that's what happens.

THE LARGEST CANNABIS PRODUCING COUNTRY

With the growing number of weed users world-wide, the demand for marijuana is at its all time high, following its legalization in several States in the US now. There are countries, such as Uruguay, that legalized cannabis while many other countries have or in the process of decriminalizing possession.

And although it is still illegal in many countries around the world, the export of weed has sustained to be a thriving industry regardless of its prohibited status.

Which Country Produces the Most Weed?

The United States, being the most mainstream in marijuana propaganda, as well as having the most media coverage about marijuana at present time can easily be assumed to be the number one country that produces the most cannabis worldwide.

Although some countries have yet to legalize marijuana in their area, profitability, tradition and religion has kept the practice of cultivating marijuana in these countries alive.

Here is a list of top Countries in the world that yield the most marijuana:

India

Marijuana use in India dates back to a thousand years ago. Used in medical, recreational and religious purposes, cannabis has always been a part of the culture.

India is said to be the motherland of cannabis and where all the different strains of marijuana originated. Marijuana in India is usually grown in the northern regions of the country's territories.

South Africa

The legal use of marijuana in Africa is prohibited, but the rich landscape and warm weather suits the widespread cultivation of cannabis. Because of an indecisive government, almost every African country has some form of marijuana cultivation, but South Africa has yield the most output. Due to the scarce police network for export, South Africa shows to have the qualities required to transport marijuana effectively worldwide.

Mexico

Mexican cartels have been responsible for a lot of drug imports that come into the southern part of the United States. Because of a high profitability after crossing the border, a lot of these drugs including cannabis are smuggled in high quantities almost on a daily basis. Border patrol have been trying to suppress the influx of drugs coming into the United States for many many years. With the increase of border protection services, the more clever those transporting the goods have had to be.

The transport methods have insured Mexican cannabis to be in abundance and can be seen everywhere throughout the southern region of the United States.

Paraguay

Mexico, having to jump borders to deliver cannabis to the United States, Paraguay has established themselves as the largest suppliers of cannabis to the rest of South America. Being ranked number 2 in the world as a top cultivator in cannabis, Paraguay supplies countries Brazil, Argentina, Uruguay and Chile in a constant basis. Despite the government's weight in enforcing a stop to this outbreak, farmers are still eager to grow the crop because of its return in value with an estimated worth of over 500 times more than most standard crops.

Afghanistan

Although foreign forces from the US and UK occupy the country following the downfall of Al Qaeda, the cultivation and export of drugs in Afghanistan has grown rapidly.

Al Qaeda, being effective in eradicating drug crops like cannabis and opium in Afghanistan, have had their presence diminished in the county, resulting to farmers growing profitable but illicit crops in huge farm lands everywhere.

Marijuana in Afghanistan can be seen growing wild in the streets around the city and the country being labeled once the producers of the finest cannabis strains in the world, it is said that Afghanistan is once again on the road to establishing this title.

Today, Afghanistan has been recorded to be producing the largest yields of cannabis per hectare in the world, producing up to 1,500 to 3,500 tons of cannabis a year.

Which Country Smokes the Most Weed?

After reviewing countries that yield the biggest marijuana crops worldwide, the consumption of cannabis can easily be assumed to go hand in hand with the top players that grow cannabis.

Some would also call out Jamaica as the home of the most smoked country in the world… but this is in fact not so.

In contrast to the above, the top countries that consume the most cannabis does not coincide with countries that yield the most crops.

New Zealand

Cannabis is widely used in all provinces across New Zealand. Cannabis is dubbed the country's third most used recreational drug following nicotine and alcohol. Even though 14.6% of the country's total population is smoking marijuana, marijuana use in New Zealand is still illegal. Efforts to remove penalties and prohibition of marijuana use is already in motion and can be ex-pected in the near future.

Italy

Italy has already legalized marijuana use. With 14.6% of the country's total population smoking marijuana as a recreational and medicinal drug. Italy's marijuana laws are only limited to personal use, the production and distribution of weed in the country is illegal till this day.

United States

The country that introduced medical marijuana to the world, the United States has a marijuana smoking population of 14.8%. More than half of the results are based on 24 states that legalize Medical Marijuana. Today, the state of Colorado is the only state that legalized marijuana in a recreational level.

Zambia

Although the drug is illegal in the country, a 17.7% statistic on Zambia's total population smoke marijuana. Because of marijuana's high profitability, a growing number of advocates import marijuana from other neighboring countries.

Iceland

According to the UNODC (United Nations Office on Drug Crime) the people that consume the most cannabis, in relation to its population, are those that reside in Iceland.

Icelandic residents that smoke marijuana accounts to 18% of its population according to a report dating back 2012.

Total population of Iceland that smoke weed is greater than the 15.4% of Americans in the United States that use the drug for recreational and medical purposes.

The study shows that there are more people in Iceland that consume cannabis regularly, while another study determines the country with the most population that consumes cannabis.

WHERE DOES MARIJUANA GROW NATURALLY IN THE WORLD?

Marijuana is a fast growing bushy plant with large sticky flowers, mostly famous for its psychoactive compounds. It can be used when dried or processed in different ways to yield a variety of products that can be eaten, smoked or vaporized. Marijuana is used all over the world is famous for recreational, medicinal and industrial purposes. It is due to its psychoactive properties that marijuana was banned in most parts of the world in the twentieth and twenty first century and remains so to this day. Even after it was declared illegal to possess or use marijuana, the popularity has not died down and it is used extensively all over the world today.

The Marijuana plant originated in the Himalayas, around Tibet. It was discovered by the Chinese who used it for a number of things. But soon after, the Chinese discovered another drug, called opium and marijuana gradually lost its flavor for them. However, marijuana seeds were scattered in the region and the plant found its way to India. The Indians used marijuana for making ropes and oils. After that, the Arabs spread the plant to Afghanistan, where it still grows in huge amounts and some plants are as old as a thousand years.

Marijuana is a very easy to grow plant, if you acquire the right seeds. Marijuana grows wildly all over the world but the quality of such plants is often very low. Wildly growing marijuana is refered to as "ditch weed" because it is never as potent as marijuana that is grown intentionally.

Nowadays, marijuana is found in most parts of South Asia, especially Afghanistan. In America, there is a large production of marijuana in South America. The reason for this could be that marijuana is a plant that thrives in humid and hot environments. Even though marijuana originated in the Himalayas, which are cold and dry, the plant has evolved and adapted to much denser climates.

Basically, you can grow marijuana anywhere. It would be difficult to grow it in a desert or say, the Arctic. The desert would not only fail to provide the required amount of water, but the soil or sand would lack the necessary nutrients required for a plant like marijuana to grow properly. Extremely cold places, like the arctic would also prove to be unwelcome hosts as these places have very low climates and are usually extremely dry which makes it very difficult for a plant like marijuana to grow properly.

Marijuana grows best during summer and spring as these seasons are usually hot and provide the optimum temperature. The climate also gets quite humid which only adds to the growth of the plant. Marijuana can be grown both indoors and out-doors, it grows well in both scenarios. Because marijuana is a such an easy plant to grow, you will find it everywhere, even as a wild plant on the roadside.

WHERE DOES CANNABIS GROW IN THE WILD?

As a result of human intervention on the cultivation of marijuana, we have reached the point where we're able to grow the plant indoors. But before humans began cultivating marijuana for their own personal use, where did it grow in the wild? It grows as rampantly as any other weed in some parts of the world. Check this piece to explore where marijuana grows in the wild and how to find it.

In more recent times, marijuana has been known to grow in locations that have been specifically selected to grow it. That means that most marijuana plants have been a result of human intervention on the cultivation of marijuana, rather than actually growing naturally. The marijuana plant that we smoke today is more the product of human intervention than it is nature. The final product that reaches the consumer is usually the unpollinated female cannabis plant, and this is not what usually grows in the wild.

The kind of cannabis that grows in the wild is usually known as "ditch weed" and is usually a small, bushy type of cannabis that doesn't really contain high levels of THC. Of course, that's not to say that high-level THC marijuana does not grow in the wild, although it's unlikely plants get to that stage before being eaten by a deer. So where does marijuana grow naturally? This is an exploration into the places in the world where marijuana grows in the wild, and the potential reasons that it does so.

What kind of climates does marijuana like?

It's important to consider the kinds of climates marijuana likes, as this can give some insight into where it grows naturally. In general, marijuana likes warmer, more humid climates. That doesn't mean it doesn't grow up in the mountains, of course. There are people cultivating marijuana high up in the Himalayas in Nepal. Marijuana is quite a resilient plant, and it can grow basically anywhere except for the desert and Antarctica. This is one of the reasons that marijuana has been able to spread its seed across the four corners of the globe. It just requires enough sun and a little bit of heat, and it's good to grow.

Marijuana originated near the Himalayas near Afghanistan, so it is known to grow naturally in drier, cooler climates. However, it has since then spread all over the world through human beings' very keen interest in it. Through cultivating it all over the world, we have found that marijuana grows in a range of different climates and can survive just about anywhere in the world. These days, we see that marijuana grows better in humid, warmer climates, so it is not unlikely to find marijuana in places far from where it originated.

Where are you likely to find wild cannabis?

In general, it is safe to say that there are probably wild cannabis plants growing all over the world. A lot of them may be unrecognizable to most people because it is a smaller, bushier hemp plant. However, there are known to be fields of marijuana plants in certain parts of the world, because their climate allows the plant to grow quickly and strongly. In general, those places which experience warmer, more tropical climates are the home of most of the wild weed growing in the world.

There are known to be enormous fields where marijuana is sprouting left, right, and centre in Mexico and Jamaica. Marijuana has been an enormous part of the culture in Jamaica because of how readily available it is there, and even the laws in Jamaica are finally starting to catch up. The heat and the humidity in these two parts of the world mean that someone just has to basically throw a seed into a field, and you can pretty much guarantee that it will grow into a marijuana plant.

Marijuana is also said to have been growing naturally in the Midwest of the USA for decades now. These plants are probably remnants of what used to be an enormous hemp culture in the USA. Once upon a time in American history, it was considered patriotic to grow hemp because of how important it was to America's agriculture. Because marijuana is a weed, it has the characteristic of sticking around in a particular location for a very long time. However, this marijuana is usually quite low in THC and so doesn't get the user high the same way commercial-grade marijuana might. Marijuana is cultivated these days with the very purpose of having a very high THC content, but marijuana that grows naturally does not necessarily have the same intentions.

When is a good time to go wild weed hunting?

So, it doesn't really matter where you live in the world — chances are there is wild weed growing somewhere. It is less likely to be growing in the parks of major cities than to be in the mountain areas, but it is certainly growing in parts of Africa, Asia, and the USA. So, when is a good time to go looking for weed that is growing in the wild?

Those who are growing outdoors know that most marijuana plants are flowering in the autumn time of the year. Of course, a lot of wild marijuana plants do not necessarily make it to flowering time because of predators such as deer. But if you do manage to find a little treasure that is still alive by the fall, that is the best time to go picking. During the autumn is when the flowers of a marijuana plant begin to appear and turn into the kind of dense buds that growers like to pick. So, irrespective of where you are in the world, if there is marijuana growing there, it's time to go picking during the autumn!

Walking down the street and finding a marijuana plant growing naturally is an unusual occurrence for most people. Most marijuana is grown these days on farms, and those plants that are found growing in the wild are usually hemp plants. If you do happen to be walking in a field nearby and see a marijuana plant, it might not be wild. Some plants are being grown illegally in locations where it might seem like there is a plant growing naturally. So beware when you are going out to look for wild marijuana plants. A group of tall plants growing somewhere where you do not see any other marijuana plants could be a sign of someone hosting an illegal operation. Be careful not to cut down any plants that could get you into trouble.

WHO IS THE WORLD'S TOP CANNABIS PRODUCER?

The United Nations Office on Drugs and Crime (UNODC) recently issued its World Drug Report 2017—its 20th annual survey of production, trafficking and eradication and enforcement efforts around the globe.

In past years, the report has sought to quantify the amount of cannabis cultivated in each producer country—over the past decade consistently placing Morocco in first place, generally followed by Mexico and Paraguay. This general trend continues—with some new variations.

The report only attempts to quantify the area under cultivation for a handful of the top producers.

Morocco is again way out in front, at 47,000 hectares under cannabis cultivation. It is followed by Mexico at 15,000 hectares. But now, relative industry newcomer Nigeria is in third place at just over 4,500 hectares. Next in line is Lebanon at 3,500. Paraguay clocks in next at just over 2,780.

There is a certain lag-time in the UNDOC reports, as they are based on most recent available figures—the 2017 report draws on on data collected between 2010 and 2015. The estimates are based on "direct indicators (cultivation or eradication of cannabis plants) or indirect indicators (seizures of cannabis plants, domestic cannabis production being indi- cated as the source of seizures, etc.)..."

A total of 135 countries are said to be producing cannabis based on these indicators.

Morocco remains the country most reported by governments as the source of seized "cannabis resin" (hashish), followed by Afghanistan and, more distantly, by Lebanon, India and Pakistan. Hashish trafficking was found to be more often "interregional" (notably, from North Africa to Europe), as opposed to trafficking in herbaceous cannabis, which is largely "intraregional" (for ex- ample, within South America).

The report acknowledged: "Measuring the extent of eradication is challenging because some countries report eradication in terms of hectares, while others report in terms of numbers of cannabis plants eradicated, weight of cannabis plants seized or number of cannabis cultivation sites eradicated. This makes comparisons of eradication difficult."

Leading the field in the prior category (area eradicated) is Mexico, followed by Morocco and Nigeria, based on 2010–2015 figures.

The largest numbers of cannabis cultivation sites eradicated were reported by the United States, followed by Ukraine, the Netherlands and Russia.

The largest numbers of cannabis plants eradicated were reported by Nigeria, followed by the United States, the Philippines and Paraguay.

The largest quantities of cannabis plants seized were reported by Bolivia and Peru, followed by Jamaica.

In terms of tons of seized "cannabis herb," the United States is in the lead, followed by Mexico,

Paraguay, Colombia and Nigeria. For tons of "cannabis resin," Spain is the global leader (stuff smuggled in from North Africa), followed by Pakistan, Morocco, Afghanistan and Algeria.

Not surprisingly, more hashish is seized in the Old World, and more herbaceous cannabis in the New World. In 2015, almost two thirds (64 percent) of the total quantity of cannabis herb seized worldwide was seized in the Americas, the report finds.

Seizures are growing fastest in Africa and Latin America—more than doubling in both these regions in the 2010-2015 period.

But, as always, it remains unclear whether this is due to increased production or stepped-up enforcement.

LEGAL CANNABIS INDUSTRY POISED FOR BIG GROWTH, IN NORTH AMERICA

According to Arcview Market Research and its research partner BDS Analytics, over the next 10 years, the legal cannabis industry will see much progress around the globe. Spending on legal cannabis worldwide is expected to hit $57 billion by 2027. The adult-use (recreational) market will cover 67% of the spending; medical marijuana will take up the remaining 33%.

The largest group of cannabis buyers will be in North America, going from $9.2 billion in 2017 to $47.3 billion a decade later. The largest growth spread, however, is predicted within the rest-of-world markets, from $52 million spent in 2017 to a projected $2.5 billion in 2027.

The worldwide adult recreational cannabis market remains hampered by the United Nations and its 1961 Single Convention on Narcotic Drugs. The Arcview and BDS report believes nothing will be done to change the U.N. attitude until U.S. federal laws legalize marijuana — something Arcview's CEO, Troy Dayton, believes will happen after the 2020 presidential election.

Still, the main difference between the U.S. and European cannabis markets is that in the U.S., recreational use will dominate sales. With a budget of $1.3 trillion in health care spending, European government-subsidized health care systems will bring the medical cannabis market to dominate Europe and become the largest medical marijuana market in the world.

Tom Adams, editor-in-chief at Arcview Market Research and principal analyst at BDS Analytics, points out that the big news in 2017 was Germany opening up cannabis for medical use in pharmacies. He celebrates a big and constant turnaround in the worldwide cannabis market ahead.

Highlights of the 65-page report "The Road Map to a $57 Billion Worldwide Market" include:

- The initial decision by many U.S. states and Canada to create medical-only cannabis regulations prompted many other countries to act similarly. But California's and Canada's willingness to legalize adult rec-reational use triggered a second wave of laws internationally to increase access to medical cannabis.

- South America has some of the most liberal medical cannabis programs. Led by Brazil, Argentina, Peru and Uruguay (the only country in the world in which adult recreational use is legal for all its citizens), the South American medical cannabis market may grow from $125 million in 2018 to $776 million in 2027.

- Germany is poised to be the leader of the European cannabis market, and Italy is expected to be second with $1.2 billion in sales by 2027. Overall, however, the European cannabis market is not expected to grow as stridently as its potential suggests.

- Australia's legal cannabis market is forecast to grow from $52 million in 2018 to $1.2 billion in 2027, the 5th largest in the world.

- Israel has a small population and a long history of legal medical marijuana use. It continues as a leader with years in the development of cannabis pharmaceuticals.

- Canada is among the few countries where investors have already shown confidence in the future legality of the cannabis industry; they are betting with billions of dollars pouring into public equity investments.

CANNABIS PRODUCTION & DISTRIBUTION

Marijuana and Hashish Trade

Marijuana is grown and trafficked all over the world, while cannabis seized in the United States is either grown domestically or smuggled from Mexico or Canada. Other countries known for producing and distributing marijuana to the U.S. are Colombia, Jamaica, Kazakhstan, Thailand, South Africa, and Nigeria.

Bricks of Marijuana

Forming marijuana into compact bricks is one of the more popular ways to transport bulk amounts of marijuana across borders and within the United States.

Marijuana Plants in U.S. National Forests

Most of outdoor cannabis cultivation in the U.S. occurs on public lands where cultivators take advantage of remote areas to minimize the risk of forfeiture. More than 3 million marijuana plants, which equates to 3,000 metric tons, have been found and destroyed in National Forests in the United States since 1997. These marijuana farms have been planted in protected areas, often destroying the surrounding areas by the use of herbicides and pesticides.

Home-grown Marijuana

Home-grown horticulturists have gained knowledge in breeding by cross-pollination and nurturing a vast variety of strains. Many breeders now concentrate on developing varieties in the plant by controlling the growth process. This is done by using heating lamps, fluorescent bulbs, ventilation and soil nutrients, hydroponics (growing without soil by using a liquid solution which contains nutrients and minerals) and salt-free sand.

Marijuana Growing and Cultivation in Soil

Successful outdoor grows will depend on Mother Nature. Having sufficient water and rich soil will increase the chances of having a good crop. However, like any other crop, the grower has limited impact due to climate. Indoor plant operations are the most successful because the grower can control the growing environment. For both types of grow, the basic requirements needed are light, heat, ventilation, food, and water.

Step One

Once the seed is planted growers wait until germination has taken place. After evidence of a healthy root system begins to emerge from the base of the medium and the first leaves appear, the seedlings are ready to be transplanted or repotted.

Hydroponics

Hydroponics simply means growing without soil, using a liquid solution which contains all the nutrients and mineral required to produce a healthy plant. Many experienced growers prefer hydroponics due to the faster growth rates and larger plant yields.

Marijuana Production Harmful to the Environment

The production of marijuana can also harm the environment by contaminating waterways, destroying vegetation and wildlife habitat through the use of chemical fertilizers and pesticides. Indoor grows are also harmful because of the increased fire risk posed by rewiring or jury rigging electrical bypasses in grow houses.

THE WORLD'S BIGGEST
CANNABIS FARM

On a quiet country lane in Kent stands a short row of picturesque houses, each with its own drive and large garden.

By the side of the road, a sign points the way to a children's nursery. It could not be a more peaceful setting.

But while it might seem a run-of-the-mill scene in Middle England, the folks who call this place home have no idea that they have an exotic — not to mention controversial — neighbour.

Just a farmer's field away lies Britain's biggest cannabis factory, churning out weed with an annual street value of £80million.

Two enormous greenhouses mark the spot where 30,000 of the banned plants are grown in top-secret conditions, protected by security patrols, CCTV cameras and motion sensors.

The dozens of men and women tending the leafy plants are no gun-toting gangsters, however.

Instead, they are white-coat clad scientists whose work is all legal.

They are employees of British firm GW Pharmaceuticals, working at the country's only research facility licensed to grow cannabis on such an enormous scale.

It it believed to be the largest fully legal medical cannabis-growing operation in the world.

While the location is kept a closely guarded secret, The Sun can reveal it is situated on an anonymous industrial park.

The facility is almost three times the size of Britain's biggest known illegal cannabis-growing operation, uncovered in Bangor in 2009.

Since it opened in 1998, the factory has produced around two million cannabis plants, most of which have been used for medical research or for the production of a drug called Sativex, to help those with multiple sclerosis (MS).

MS affects nerves in the brain and spinal cord, causing problems with balance, vision and muscle movement.

Sativex was the first cannabis-based medicine to be licensed in the UK. It is an oral spray which treats the muscle stiffness which afflicts sufferers. It produces no chemical high.

While the drug is available on the NHS in Wales, it is not in other parts of Britain due to its cost, meaning many users choose to pay for it privately.

Now trials are being conducted to find out whether chemicals extracted from the plants could also be used to treat other conditions.

The firm's director of botany and cultivation, Dr David Potter, says: "We are testing childhood epilepsy drug Epidiolex in America, and the results are extremely compelling.

"We're seeing an average reduction of 50 per cent in the number of seizures and, in some cases, it is cut to zero.

Other possible uses for cannabis-based medications could include the treatment of schizophrenia and diabetes, while there are hopes Sativex may be approved to ease pain in terminal cancer patients.

As a result, GW Pharmaceuticals is expanding its already-huge glasshouse facilities in order to cope with the booming demand for cannabis for medical research.

Shockingly, despite the size of the facility and its scientific importance, only a handful of the locals we speak to know about the growing operation on their doorstep.

Retired Glynis Chatfield, 62, says: "I'm shocked it is located here.

"I've heard some people with MS use cannabis because it helps with their symptoms but this is a much better way of doing things.

"Even so, it's not the kind of thing you expect on your doorstep."

The factory setting, complete with fans, bright lights, and precisely-controlled levels of fertilisers, is necessary in order to produce a uniform medical product.

Unlike street cannabis, these plants do not contain any dangerous contaminants or toxic residues.

Despite the quiet location, the facility is ringed with high-security fencing to keep out criminals.

Cannabis is classified as a Class B drug, with a maximum sentence of 14 years for those convicted of producing the drug.

In fact, tough Home Office regulations meant the firm had to spend five months beefing up site security before it could open.

Should any unwanted visitors make it inside, motion and temperature sensors are primed to alert staff, while every corridor is kept under continual video surveillance.

One local who used to work at the plant but did not wish to be named, reveals that employees are also subject to close observation. She says: "There are cameras everywhere — and you have to sign in with a security officer.

"The greenhouses are huge and crammed with cannabis plants. To start with it was strange working with the plants around but after a while you got used to it."

The firm's director of botany, Dr Potter, adds: "Once you're in, it's more like a garden centre. It's a very serene working atmosphere."

Despite being surrounded by the plants, Dr Potter says he's never been tempted to experiment with the drug, joking: "Real ale is my drug of choice."

Since the plant opened, there has never been a major security incident. But any drug dealers keen to get hold of the stash might end up disappointed, as around half of the 20 tonnes of dry weight cannabis grown here every year would produce no high in recreational users.

That's because rather than THC — an ingredient that makes users "stoned" — the plants have been bred to be high in other chemicals, including

CBD, currently being researched as a possible anti-schizophrenia medication.

Dr Potter explains: "The thrust with recreational cannabis has been to select seeds from the plants which have high THC levels.

But we've gone back to basics, and have plants not just rich in THC, but other cannabinoids — chemicals unique to cannabis."

This range of different chemicals is the reason cannabis-based medicines can be used in treating a variety of illnesses. And while some of the medicines may contain the active ingredient of THC, the addition of CBD prevents it from having an intoxicating effect.

Meanwhile, high levels of THC in modern "skunk" cannabis — estimated to be twice as potent as that available in the Sixties — are threatening to create a mental health disaster for street users.

High levels of THC can induce temporary schizophrenia-like symptoms such as paranoia, delusions, anxiety and hallucinations, with some research suggesting it can cause long-term mental problems.

While CBD chemicals can protect against these effects, it has almost vanished from modern cannabis due to dealers' selective breeding.

Dr Potter explains: "The THC content has gone up, and the CBD content has gone down. There's been an increase in the psychoactive ingredient, and the one that was anti-psychotic, mother nature's antidote, has disappeared.

So you increase your risk twofold. The actual threat to mental health in the young especially is a concern. It's not a soft drug."

Crime Prevention Minister Lynne Featherstone says: "The law allows for a licence to be issued, under strict regulations, to companies growing controlled substances such as cannabis for research and medicinal purposes.

Decisions are made with proper regard for management of risk and security."

A government spokesman adds: "The UK does not recognise that cannabis, in its raw form, has a medicinal use. There are people with debilitating illnesses who may not find relief from existing medication. For them, the UK does recognise the medicinal value of cannabis-based medicine Sativex.

TOP 5 CANNABIS STRAINS FOR OUTDOOR GROWING

If you haven't thought of an outdoor grow for this season, it's about time to do so.

Outdoor growing is an extremely rewarding process and usually worth it due to a very beneficial ratio between initial costs and final yields. Many of Royal Queen Seeds' varieties show very good results when they're grown naturally under direct sunlight.

However, the particular strain you'll choose does have an impact and can determine success or failure. We decided to put together a Top 5 of our strains that are suitable to be grown outdoors and finish by the end of September at the latest. By

choosing either one of these strains, you'll have high chanced to avoid unnecessary trouble with autumn rains in October. While other growers are still busy inspecting their buds for mould, you're already done with the entire process and filled up all your curing jars. If you missed out on our Top 5 strains for colder and warmer climates, feel free to give those a quick look as well.

Top 5 Cannabis Strains For Colder Climates

Top 5 Cannabis Strains For Warmer Climates

HONEY CREAM (FAST FLOWERING)

- BlueBlack x Maple Leaf Indica x White Rhino

- Indica-dominant (65%)

- Flowering period: 6-7 weeks

- Harvest month: September

Let's start off with Honey Cream Fast Flowering, one of our recently updated favourites. This variety is becoming more and more popular in Spain due to its sweet, creamy aromas but is suitable to be grown anywhere in Europe. The flowering period is remarkably short with 6-7 weeks allowing the cultivation in regions where climatic conditions shift rather early. However, July and August usually provide a sufficient time window for Honey Cream to receive a lot of sunlight, giving growers the chance to harvest good amounts of sweet-fragranced, potent buds in September.

Besides a short flowering period and amazing flavours, Honey Cream convinces with a solid Indica background consisting of BlueBlack (Blueberry x Black Domina), Maple Leaf Indica (100% Afghan Indica), and White Rhino. These three strains give Honey Cream mostly Indica genetics, but there's also a decent amount of Sativa potency left from White Widow, the predecessor of White Rhino. When things go well, plants reach outdoor heights of 180-250cm and yield up to 675g per plant.

Honey Cream outdoor

SPEEDY CHILE (FAST FLOWERING)

- Green Poison x Chile Indica landraces

- Indica-dominant (70%)

- Flowering Period: 6 weeks

- Harvest month: early September

Speedy Chile is the next of our Fast Flowering varieties and a very good choice for outdoor growers. This strain combines Green Poison and several Chile Indica landraces, although a third-generation autoflowering plant is partly responsible for an incredibly short flowering time of 6 weeks (42 days!). Don't be confused; Speedy

Chile does not carry the autoflowering trait and remains a feminized strain that flowers depending on the amount of light hours per day. One could say that Fast.

Flowering strains are the next breeding evolution since autoflowering genetics – these strains produce higher THC levels while maintaining an incredibly short total crop time. Outdoor growers and novice cultivators come to an excellent decision when choosing Speedy Chile Fast

Flowering for their next grow. Fully mature buds can be harvested in early September, leaving not the tiniest bit of room for uncertainty regarding climate or mould issues.

Speedy Chile outdoor

CRITICAL KUSH

- Critical x OG Kush

- Indica-dominant (80%)

- Flowering Period: 7 weeks

- Harvest month: late September

You've probably already come across Critical in your career as a hobby grower, a true heavy-yielding champion we recommend for indoor and outdoor growers seeking one of the best ratios between yield and flowering time. Critical, which also made it into our Top 5 For Colder Climates, is a vigorous and rapidly flowering Indica that deserves to be in every single ranking on cannabis strains that exist. However, crossing Critical with the notorious OG Kush took potency to a whole new level.

Critical Kush features some of the cerebral and rather Sativa-like effects of OG Kush, in addition to the powerful couch-lock stone of Critical – a classic scenario of "buy one high, get one free". Growers looking for a sturdy, potent, and heavy-yielding strain for the outdoors won't make a bad decision with Critical Kush. The rewards you'll receive in late September are huge; Outdoor yields of up to 550g per plant, long-lasting effects of sheer intensity, and a final product with average THC levels of 20%.

Critical Kush outdoor

SPECIAL QUEEN 1

- Power Bud x Skunk

- Indica/Sativa (50/50)

- Flowering Period: 7-8 weeks

- Harvest month: late September

Special Queen 1 is Royal Queen Seeds' classic Skunk. It's suitable to be grown outdoors in many different environments, also in colder climates of Northern Europe.

Novice growers among our customers achieve very good results with this balanced Indica/Sativa hybrid, just like experts who seek a reliable, stable, and competitively spiced Skunk variety. Special Queen displays great vigour throughout the entire season and grows huge outdoors, sometimes exceeding outdoor heights of 220-270cm in good conditions. Outdoor yields of up to

550g per plant will be the reward for growing this queen, and she'll be ready for harvest in late September.

The decision to grow Special Queen is not a hard one to take. This strain is one of the best options for growers who want to reduce the risk of choosing a strain that is not robust enough for the outdoors. Skunk laid the foundation for a large percentage of today's varieties for a reason – it's simply one of the best and most stable strains that ever existed.

Special Queen

- Black Domina x Kalijah

- Indica-dominant (85%)

- Flowering Period: 7-8 weeks

- Harvest month: late September / early October

Royal Domina is a potent Indica masterpiece created by crossing Black Domina with the exotic genetics of Kalijah. This strain stays comparatively short outdoors, growing into average outdoor heights of 140-180cm. At the same time, Royal Domina yields up to 600g per plant and can be harvested by late September to early October, depending on the climatic conditions throughout the season. Additional key features of our Royal Domina are enticingly sweet aromas, an impressive resin production, and high THC levels of 20%. This strain is perfectly suitable for Indica lovers and will be an excellent choice for anyone who desires to harvest large, heavily fragranced blossoms with a thick layer or frosty resin. Many growers know about the qualities of the infamous Black Domina but we believe we've created something even better – Royal Domina is more classy, has an improved flavour profile, and is at least equally good at forcing bodies and minds into a state of deep relaxation.

TOP 7 LARGEST LEGAL MARIJUANA GROW OPERATIONS IN THE WORLD

The prevalence of legal marijuana for both medical and recreational use has only grown over the years. With the increased demand, several LARGE grow operations have been set up legally throughout the world to provide the best bud for mass consumption.

Tweed Headquarters, Niagara-on-the-Lake, Canada

This 350,000 square foot growing facility in Southern Ontario is home to a variety of carefully selected cannabis species. The state of the art facility has been producing products for Tweed Farms, one of the largest companies in legal marijuana, since 2014. Home to the Canadian strains of Snoop Dogg's personal brand, Leafs by Snoop. This facility oversees and tracks the growing of thousands of plants with a systematic tagging process that completely keeps track of every step and movement of each plant. The partnership with Snoop Dogg and long standing reputation of quality MMJ products makes this facility even more impressive. The carefully crafted crops yielded through this grow operation are currently available to MMJ patients in Canada only.

Bright Green Group of Companies, Western, New Mexico

Bright Green is currently in the process of beginning construction on what is planned to be the biggest medical marijuana grow operation in the United States. The Delaware based company intends to grow mainly marijuana along with other homeopathic plants which oil can be extracted from. The new facility is looking to be around six million square feet and placed on native American tribal lands. The agreement with the Acoma Pueblo tribe will allow for the company to only "have to" operate under tribal and federal laws, leaving out the necessity of a state license through New Mexico. This large-scale project is aimed at looking towards the future of the marijuana industry and being able to handle the capacity of marijuana production needed for future recreational use and medical use as well.

Americann, Freetown, Massachusetts

This canna business park scheduled to break ground in 2017 will feature fifty-three acres and one million square feet of marijuana growth and production areas. With Massachusetts recently voting recreational marijuana into place it is the perfect time for such a large-scale operation to open up for business. The project is aiming for energy efficient marijuana production and large scale quantities to supply dispensaries with marijuana for both recreational and medicinal uses. The business park will be hosted by Americann and allow other investors and businesses to lease space for legal marijuana production. The Colorado based parent company, Americann, is looking to expand its horizons and help bring knowledge from the recreational legalizations in Colorado in 2012 out East.

Desert Hot Springs, California

This sleepy desert town in the Coachella Valley has recently come into the spotlight. In 2014, the town opened up to the idea of legal marijuana grow operations and was flooded with offers from a plethora of professionals in the industry. Several large-scale grow operations are in the process of being built. The main problem with this location is the distance from serious civilization. Getting the amount of electricity and water to the area has caused problems for buildings that need to draw enough electricity to power a small town. Still, investors like GFarma and Canndescent, who just raised 6.5 million dollars to fund their Desert Hot Springs facility, are building grow operations made up of thousands of square feet to help supply the growing marijuana demand. Since the legalization of recreational marijuana in California there will be much more of a need for product, in the already busy medical marijuana market. Desert Hot Springs isn't the only small town searching for added revenue through marijuana growth through, dozens of other California towns are opening their free space to industrial zoned marijuana grow operations areas as well. Keep an eye on this small I-10 pit stop in the next year for openings of many big-time operations.

Kannaswiss, Switzerland

One of the largest grow operations in Europe lies in Switzerland. This six year old facility was opened to produce the low potency, legal cannabis for Swiss connoisseurs. The owners of the grow operation focus on oil products that they extract from their three thousand plants, but also supplies flowers. Kannaswiss produces their marijuana products of one percent or less THC content in a country pot barn in their native Switzerland. The legal marijuana industry is still blossoming across Europe, but countries like

Spain and Switzerland are now participating in recreational use of the plant.

Harborside Farms, Salinas Valley, California

Harborside Farms is a forty-seven-acre cannabis farm in Northern California. The perfect growing climate is found in this valley that until recently has been home to produce farmers and cut flower growers. The owners of Harborside took advantage of the old flower greenhouses and expanded their project to cover the acreage with hundreds of cannabis growing greenhouses. This massive operation boasts 360,000 square feet of growing space and an impressive 100,000 plants. With fierce competition in the California market the veteran owners of Harborside are looking to scale quickly and be able to produce mass amount of marijuana in their new facility.

aya Foundation, Columba, Chile

Located just outside of the capital city of Santiago, the Daya Foundation has successfully set up the largest legal grow operation in Latin America. The purpose of this grow op is to produce medicinal marijuana for patients across the country. In 2015, the country deemed medical marijuana to be legal and available for patients in need. Daya got to work on their farm and came up with a facility to help the masses. The current farm is looking to yield 1.65 tons of harvested marijuana this spring and will be able to help around 4,000 patients. So far, Daya is the main producer of legal cannabis in the country and plans to continue to serve patients through the new facility. The foundation provides education and access to many forms of supplemental alternative treatments including, reiki and acupuncture as well as marijuana. With Latin America now throwing marijuana into the conversation more often now, we could see more from this company in the near future.

HOW TO GROW CANNABIS OUTDOORS

The whole process of growing cannabis outdoors, from seed to flower, can be a very rewarding experience. There are certainly challenges to growing in the great outdoors and it is also true that cannabis is exceptionally hardy. The old saying that with water and sunshine cannabis will grow on a rock is quite true – to which anyone who has seen weed growing wild in Morocco will testify.

MASTERING THE ART OF GROWING HIGH-QUALITY ORGANIC MARIJUANA

If you want to grow world-class organic mariju-ana things are a bit more complicated than just lobbing a few seeds into a patch of garden and letting nature do the rest. Some loving care and attention over the months, tornadoes, floods, droughts, plagues and alien invasion aside, will ensure a fine yield of high-quality cannabis for you to enjoy. Just like homegrown vegetables, it just tastes better.

The annual life cycle of the cannabis plant starts in early spring, after the equinox, when the sun has warmed the soil and daylight begins to last longer than twelve hours. An old gardener's folk method of finding if your soil is warm enough for spring planting is if you can sit comfortably on your soil for one minute with your bare bottom you are ready to go – although you don't need to do this, it gives you an idea of the ideal sowing conditions.

These ideal conditions make seeds germinate that is followed immediately by rapid vegetation. The season has ended when mature unfertilized flowers are harvested, as the weather cools and the days get shorter. Depending on species and geographical location, generally during autumn and waning into winter.

Grow of cannabis plant

GOOD SOIL AND HEALTHY ROOTS ARE THE FOUNDATIONS OF YOUR GROW

Soil will be the anchor for your plant's healthy life. Time or money spent on good soil will provide several benefits for your plants at every stage of their growth. Healthy, bio active soil not only provides all the nutrients your cannabis plants will need for their whole life, but good soil will help control several other variables in the growth of your plants.

The following are all problems that can be largely avoided when growing in active, high-quality soil:

- Ph fluctuations

- pest resistance

- Waterlogging

- Biological attack

- Heat stress

- Fungal problems

- Nutrient lockout

Growing in soil provides a greater margin of error across the boards when there are fluctuations in any of the factors that can affect plant growth.

If you are a keen gardener and are adding cannabis to your repertoire of plants, then you already know the importance of good soil. You have been husbanding a high-quality humus over time, and your garden is rich with composts and living organisms, it is friable to the touch, no-till and holds water well while draining satisfactorily.

Making soil from several components yourself or buying a reputable quality commercial soil in bulk is another option if you are not the gardening type. A truly high-quality soil will need no fertilizers or additives for the life of your plants other than compost teas and top dressings for soil maintenance. Companion planting, mulching and throwing in a handful of worms will guarantee nitrogen fixing and soil friability, passive pest control, and water conservation.

CHOOSING YOUR PERFECT CANNABIS STRAIN FOR YOUR CLIMATE

Cannabis climate

Pot or soil. Latitude, day and season length. Recreational and or medicinal. Legal or guerrilla. Automatic or photo-period. Feminized or traditional. Mono or polyculture. Indica, sativa or hybrids. These variables you have already juggled in your decision to grow outdoor cannabis.

You already know you maybe too far north and chilly to attempt a long maturing sativa. Or your seasons are too wet and humid in general, and dense indicas can tend to rot. Perhaps the long hot days of your extended equatorial summers confuse autos that can regenerate after a brief flower period, the internal ruderalis bewildered by the seventeen perplexing hour days.

After some fascinating research, forum crawling with many truly mouth-watering bud pics an enquiring mind with literally thousands of styles of marijuana available can make a very informed decision.

It really is a kid in a candy store stuff right now on the internet. Variety truly is the spice of life when it comes to cannabis, and like many natural substances, it is a good idea to change things up to avoid building a tolerance.

It is certainly wise to grow a few species of equal strength but varying effect and have something different every day of the week. This will guarantee that different neurons are tickled by different terpenes and such avoiding strain stagnation (in our humble opinion – one plant of a few varieties is enough for a varied personal store).

MAKE SURE YOU START GERMINATING THE SEEDS IN TIME

As mentioned, the cannabis lifecycle begins in early spring, so you need to get your seeds germinating for then.

Start growing cannabis

Hours of bud research have paid off, and you have selected the beauties you wish to see in full bloom in person. Whose effects you would like to appreciate, whose aroma you would like to savour. Whose growth and flower time suit where you dwell and grow.

During germination, the seed first absorbs water through its husk by imbibition - which means to imbibe or drink. The water hydrates existing enzymes and food supplies causing the seed to swell and expand. As metabolism gets stronger hydrated enzymes become active increasing energy production for the growth process. At the same time, water increases turgor pressure encouraging cell expansion.

The first indication of life will be the cracking of the seed coat and the emergence of a small white shoot called a radical. This quickly lengthens and becomes the tap root. The new tap root pushes down into the grow medium anchoring the plant in place and begins to absorb water and nutrients. Simultaneously the new stalk reaches towards the light and leaves begin to form.

cannabis seed cracking

The first leaves to appear are oblate, thick and rubbery and are not really leaves. They are called cotyledons and are pre-formed inside the seed. When hydrated they swell considerably and are used to split the seed husk apart and protect the first true serrated leaves as the crown is forced up and outwards through the medium.

Soon a radical transformation happens called photomorphogenesis. This light dependant process makes the plant become green and begin photosynthesis. The first true serrated leaves are exposed to the sun, and vegetation has begun.

Cannabis growing is an art rather than a linear a, b, c, system. Every action has a reaction, and you will discover what suits you over time as you become a master of the alchemical flux of marijuana magic. This starts with a choice of germination method.

The best and easiest way to germinate cannabis seeds is as nature intended – in soil. Plant the seed about 0,5cm deep and cover lightly. Keep the soil around 20°C and ensure the environment is humid. The soil for germination doesn't need to be nutrient rich – in fact, a high nutrient soil will overwhelm the cannabis in this fragile phase of life. The seed has everything it needs in it to get started.

Many growers like to start their plants indoors, in a pot, where conditions are easy to control. When strong enough, plants are then hardened off, before being moved permanently outdoors – it gives the plants a strong start, and makes it less likely they will succumb to the perils of outdoor growing before they are strong enough to deal with them.

CHOOSING THE RIGHT SPOT FOR YOUR CANNABIS PLANT

Cannabis growing in the sun

Seek out a spot that is exposed to as much sun as possible. Encourage wind and rain exposure as much as possible. Rain for growth boosts thanks to the carbon dioxide dissolved in the water. Wind because a good physical stressing makes for strong plants – a larger root base to compensate for wind stress - that produce more flowers.

For a decent sized cannabis plant in the ground, a minimum of five square metres is needed per plant, or as big a pot as your space can take to provide enough root room for complete canopy and flower development.

Planting cannabis too close together forces the plants to respond by reducing side branching and increasing stretch and height. Rather than having multiple flower sites over a large bushy plant, the plant will develop one long central cola and resemble industrial hemp in structure.

Flower density is affected by planting distance as well. Well separated plants develop much thicker buds than closely placed weed and are generally less susceptible to disease and infestation as plenty of air movement is possible.

GROWING TECHNIQUES FOR GROWING CANNABIS OUTDOORS

Over the ensuing months, your cannabis will respond to regular watering and plenty of sun with vigorous growth. You will be amazed by the stretch over the full moon period or the astounding gains in volume after a summer rain. Au natural, or topping and shucking into mainlines, fimming, super cropping or low-stress training, are all grow styles proven to grow high yields of potent ganja.

low stress training

During vegetation, the plant consumes nutrients through the roots and uses light, water, and carbon dioxide as part of photosynthesis to grow as much as possible in several different ways.

- The plant gets taller.

- The leaves get larger and far more numerous.

- Side branching begins which gives the plant volume.

- The root system gets larger.

- The trunk and branches become thicker and stronger and in some instances become fluted or ribbed.

- Large knuckles form at branch nodes.

- The true genetics of your plants will express themselves. As in the large thin palmate leaves and overall branchy stretchiness of a sativa or the broad fat leaves, minimal branching and stoutness of an indica.

Be prepared with plenty of stakes and ties or netting and wire to provide support as the plants grow. Vegetating plants rarely break, and yours have been staked from the start and aren't leaning, only reaching for lumens.

Providing support as the plants grow is in anticipation of heavy flower clusters in the final weeks that can make whole plants collapse or large sections of branch snap during unpleasant weather.

Being caught out during flower time, having to run around madly fudging ad hoc support to twisting branches and leaning trees can mar delicate flowers. This is difficult work late in the game and you risk damaging the plant further having to handle it so much – so plan ahead with support.

Snake oil salesmen will entice you to buy growth boosters and vegetation formulas, but healthy soil to begin with and monthly organic top dressing is all plants require.

Vegetating cannabis plants, as do all plants, respond with vitality to monthly organic top dressing. There are a number of commercial products available that work equally well. Feather meals and fermented compost teas that contain active microbial life, bird and bat guanos or worm castings are all excellent sources of trace elements, vitamins, and carbohydrates. The rule, of course, is to err on the side of too little, burning and poisoning are still possible with organics.

WHAT IF YOU ENCOUNTER MOULD IN YOUR BUDS?

Something has gone awry in your little cannabis ecosystem. Don't freak out that your weed isn't the high definition, high rez, picture perfect weed like you see indoors.

mould on cannabis

Keep a keen eye out for moulds. Dense flower clusters can retain water that can cause botrytis or powdery mildew if there is not sufficient air flow. Tend your flowers well, remove dead and dry leaves as they can rot and moulder and spread to the buds. Remove desiccated and damaged bud material for the same reason. If you do find bud rot, remove the whole flower cluster immediately and put in a plastic bag. Try not to let any spores get on the air than can affect surrounding plants. Dispose of well or burn.

Spiders making little nests are good as they eat mites, the occasional folded over leaf where some random larvae have curled up to pupate is nothing to lose your cool about. Lady beetles are a welcome sight as are many beneficial critters drawn to your garden by its attractive vigour, and variety of species.

spider mite cannabis

Plant diseases are rare in a well setup garden. Plenty of room between plants, lots of air movement, sun, sun and more sun, not too wet, all bolstered by the renowned natural resistance of cannabis to pests, fungus, and microbial attacks should have you growing trouble free.

Seek organic solutions to pre-emptive pest control as part of the regular maintenance of your plants. Caterpillars and aphids among many other critters are discouraged by regular application of neem oil for example. Preventing infestation is far more desirable than getting rid of infestation.

THE FIRST SIGNS OF THE FLOWERING STAGE

As the days start to shorten towards the equinox, noticeable changes will occur in your cannabis plants.

During the shortening days towards the equinox and the last weeks of the growing season, but before dropping below the twelve-hour photoperiod required for full flowering, cannabis will differentiate. The growth pattern of your plants begins to alter.

Replacing the striving, stretching symmetry of vegetation with growth that begins to zig-zag and compress with less distance between nodes. The tips at the ends of branches will turn upwards creating nooks and crannies where flower formations will be cradled. Individual branches become distinguishable from the generic canopy of green.

Flowering cannabis

Quick on the heels of differentiation, proper flowering begins.

Flower clusters start to form, and the compressed zig-zag structure begins to stretch, sometimes another fifty percent of the plant's height. Brand new calyxes form in the supporting intersections of leaves and zig-zagged stalks. Turgid and already resinous pistils extend from each calyx giving the cluster the look of a tiny anenome.

The puffballs of calyxes extend along their own delicate prong, making more room for more clusters to form. Bud specific leaves start to emerge that are different to sugar leaves. They are smaller, thicker, look felty are very ridged and covered in trichomes, eventually becoming mostly submerged by the flower clusters as they swell.

Each calyx node along the protrusion will produce more calyx clusters which stack in a pattern similar to cereals like wheat or barley. Each with trichome covered twin pistils these fresh calyx florets stack one atop the other until peak fluorescence is reached. This is often when great cannabis pics are taken. The gnurled and knobbly flower clusters have a halo of pistils reaching for pollen that will never come.

This is where the fun stuff really begins. Cannabis in the raw of nature would have been well fertilized by now and thoroughly on the way to producing mature seeds. The lack of male pollen tricks the cannabis plant into producing more flowers than would be possible in a mixed sex crop in the wild. After peak blooming has been reached, the plant continues to mature, and resins are produced in copious amounts.

Microscope Resin guard

Using your choice of magnifying apparatus, a loupe or kiddies microscope for example, you can check the swelling of the resin bearing trichomes. The calyxes themselves also swell substantially, undergoing a false pregnancy, filling the unfertilized seed chamber with oils.

GETTING READY TO HARVEST YOUR PLANTS

Carpeted in trichomes that continue to swell the pistils begin to shrivel and change colour, their pollen gathering days well and truly past. Tones and shades that cover the spectrum can appear. Russet, lavender, deep brown or even hints of blue or silver, as many colours as there are strains of marijuana.

The maturing process will also see your plants morph in colours as the season comes to a close. Sugar leaves begin to mimic deciduous forests in colours and flower clusters are swollen and very firm to the touch. The bouquet of your plants will be in overdrive right now. Complex fragrances easily distinguishable from species to species are enticing and hint at the flavours to come.

trichomes cannabis

In these last weeks, the trichomes and their resin sacks begin to change colour in waves all over the plant. Usually starting with the oldest growth first.

Trichomes initially become milky rather than clear; then milky becomes a deepening amber. Ideally, you will be harvesting when the dusting of trichome colours is half white and half amber. This guarantees a peak THC content, too much longer and the THC begins to turn into other less desirable cannabinoids.

It is time to grab your favourite scissors or snips and harvest the results of your hard work.

CUTTING YOUR PLANTS

Sometime in early October for indicas and some weeks later for sativas, the trichome colours, and fragrant bouquets have inferred that it is time for harvest.

Stand back just once more and admire your handiwork before having at them.

While the plant is still standing start by removing all the leaves that have an easily accessible stalk to snip. Sugar leaves especially. This is also easily enough done with your fingernails. When done your plants will be stalks and flowers with only difficult to access leaves attached.

trimming cannabis

You have scraped your trimmers and fingers many, many times and are the proud owner of a lovely ball of dark resinous charas hashish. Consume now as you consider the next stage. If you are new to cannabis growing you will be needing a break, being amazed at how much work trimming really is.

Now break the plant down. There are no rules simply consider your drying method. Hanging a complete plant or individual long branches and detail trimming when dry. Detail trimming wet flowers and drying on screens or in a humidity controlled cupboard. Each quite valid and when done correctly provide great quality dried flowers.

As a side note, trim can be used to make cannabis-infused foods – although it doesn't have as much cannabinoid content as flowers, it still has some. Check out our recipe section to put it to good use!

Six weeks later you will be enjoying perfectly dried and cured examples of your horticultural skills.

CLOSING

Cannabis is derived from the cannabis plant (cannabis sativa). It grows wild in many of the tropical and temperate areas of the world. It can be grown in almost any climate, and is increasingly cultivated by means of indoor hydroponic technology.

The main active ingredient in cannabis is called delta-9 tetrahydro-cannabinol, commonly known as THC. This is the part of the plant that gives the "high." There is a wide range of THC potency between cannabis products.

Cannabis is used in three main forms: marijuana, hashish and hash oil. Marijuana is made from dried flowers and leaves of the cannabis plant. It is the least potent of all the cannabis products and is usually smoked or made into edible products like cookies or brownies (see Factsheet: Marijuana Edibles). Hashish is made from the resin a secreted gum of the cannabis plant. It is dried and pressed into small blocks and smoked. It can also be added to food and eaten. Hash oil, the most potent cannabis product, is a thick oil obtained from hashish. It is also smoked.

Thank you for your support. If you found any value in this book would you please do me the honor of leaving a review on the Amazon page. Thank you very much and God bless you.

```
* 9 7 8 0 9 7 0 9 7 0 1 3 8 *
```